SUN PUZZLES

by Ellen Johnston McHenry

This book was inspired by a video by Barry J. Setterfield titled "**Sun Puzzles**" (available on YouTube). Barry writes about both plasma astronomy and S.E.D. physics, and is the author of a cosmological model that combines both branches of science (*Cosmology and the Zero Point Energy* (ISBN 978-1-30419508-1).

Additional inspiration and information came from books and videos by Donald E. Scott, PhD, a professor of electrical engineering. His books are: *The Electric Sky* and its updated version *The Interconnected Cosmos*. (ISBN 979-8-9851181-0-0).

Setterfield and Scott cite the research and writings of:

Kristian Birkeland (1867-1017) Norwegian physicist, nominated for the
                                     Nobel Prize 7 times

Hannes Alfvén (1908-1995) Swedish physicist, Nobel Prize in physics 1970

Irving Langmuir (1881-1957) Nobel Prize in chemistry, 1932

Ralph Juergens (1924-1979) American engineer who was the first to write about
                                        "the electric sun"

Anthony Perrat (1940- present) American PhD electrical engineer and plasma
scientist at Lawrence Livermore National Lab and Los Alamos National Lab

Illustrations are all public domain or the author's own work, except for the Creative Commons images which have captions giving their attributions. The illustrations on pages 22, 23, and 25 are based on graphs shown by Donald Scott, which are in turn based on original work by Ralph Juergens.

Retailers can order this book from IngramContent.com

The black circle is the moon covering the sun. The brilliant light is the sun's corona. "Corona" is Latin for "crown."

The author's quick sketch of the 2024 total solar eclipse. The tiny pink dot is probably a solar flare. Solar eclipses look like this when seen "live."

Something unusual happens during a solar eclipse—we get a rare glimpse of one of the layers of the sun that we don't normally see. If you were fortunate enough to live close to "totality" during one of the recent solar eclipses, you saw the sun's *corona*. The corona will be the subject of one of our sun puzzles, but first, how does an eclipse happen?

A solar eclipse occurs when the moon comes between the sun and the earth. Of course, the sun, moon and earth are always moving, so the total eclipse only lasts a few minutes. If you are outside the area of "totality" you will see a partial eclipse, where only part of the sun is covered (which is still pretty impressive!). Solar eclipses are only possible because the sun is exactly 400 times wider than the moon, but it is also 400 times farther away. Although the sun is actually much, much larger than the moon, they both appear the same size when we see them in the sky.

TOTAL SOLAR ECLIPSE

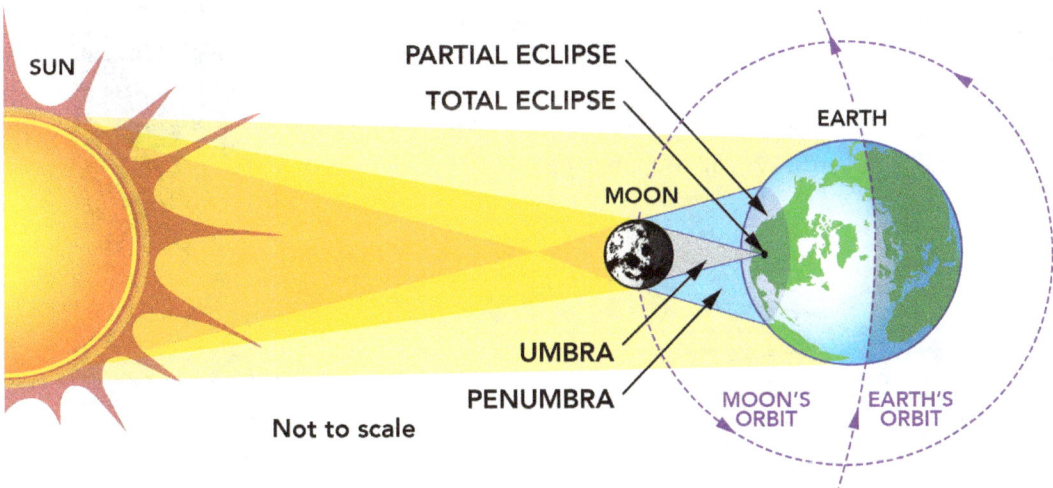

SUN

PARTIAL ECLIPSE

TOTAL ECLIPSE

EARTH

MOON

UMBRA

PENUMBRA

MOON'S ORBIT

EARTH'S ORBIT

Not to scale

Credits: NASA's Goddard Space Flight Center | https://mynasadata.larc.nasa.gov/sites/default/files/inline-images/5661_Total_Solar_Eclipse_Shadows_0.jpeg

The first sun puzzle is about the temperature of the corona. But what is the corona? Let's start by looking at diagram of what most scientists believe about the structure of the sun.

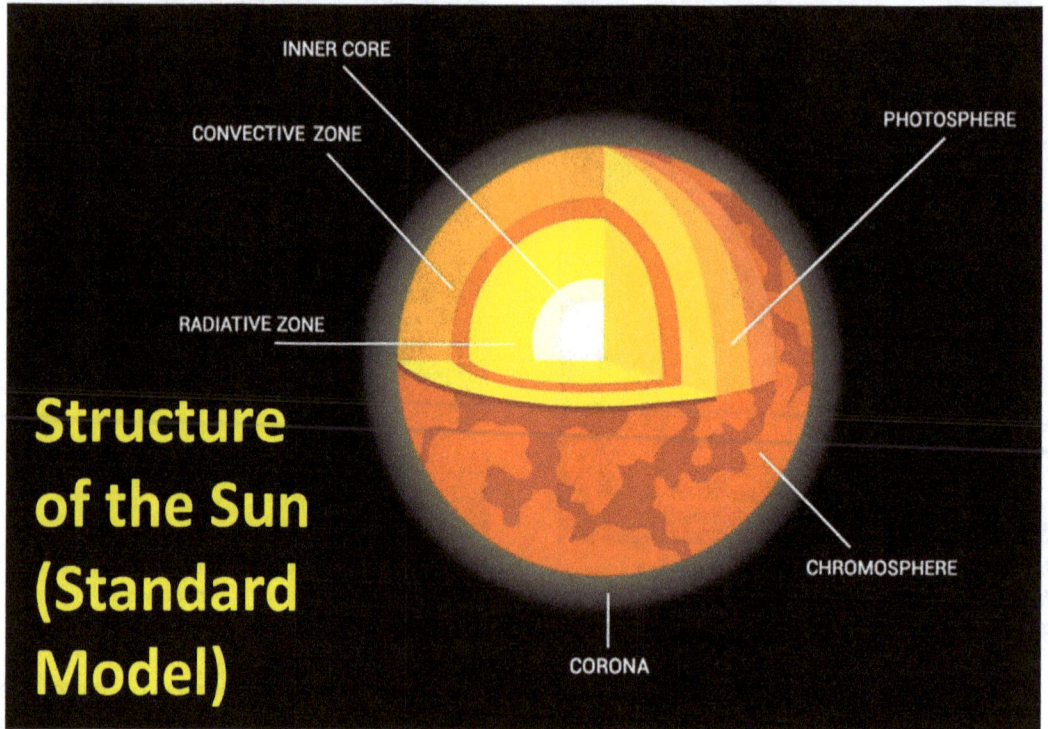

Structure of the Sun (Standard Model)

Diagram labels: INNER CORE, CONVECTIVE ZONE, RADIATIVE ZONE, PHOTOSPHERE, CHROMOSPHERE, CORONA

The corona, shown here in gray, is the "atmosphere" of the sun. The sun's atmosphere is very different from the earth's atmosphere. Here on earth, our atmosphere is made of the air we breathe, plus all the water vapor that circulates in our weather system, and then some very thin air above that. The top layer of our atmosphere, the *exosphere* lies about 200 to 600 kilometers above the surface. Our weather occurs at the very bottom of our atmosphere, in the tropo-sphere, from 0 to 15 kilometers above the surface.

The *corona* stretches for millions of miles and is about 1,000,000° Celsius. At times it can soar to 2,000,000° C. It is so hot that atoms cannot exist here—they break down into the particles they are made of, mostly electrons and protons, but also some alpha particles (two protons bound to two neutrons), and small amounts of nuclei from the atoms of the lighter elements such as carbon, oxygen, nitrogen, sulfur and magnesium. There is a special name for a mixture of these "broken pieces" of atoms (though they are not really broken!). We call this a *plasma*.

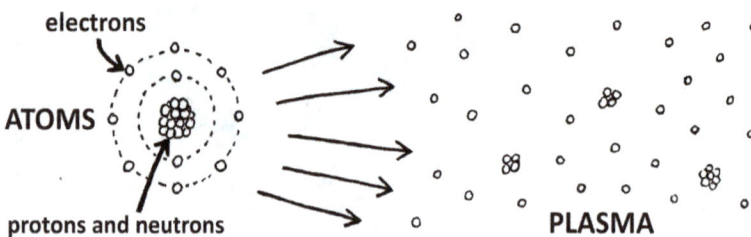

electrons
ATOMS
protons and neutrons
PLASMA

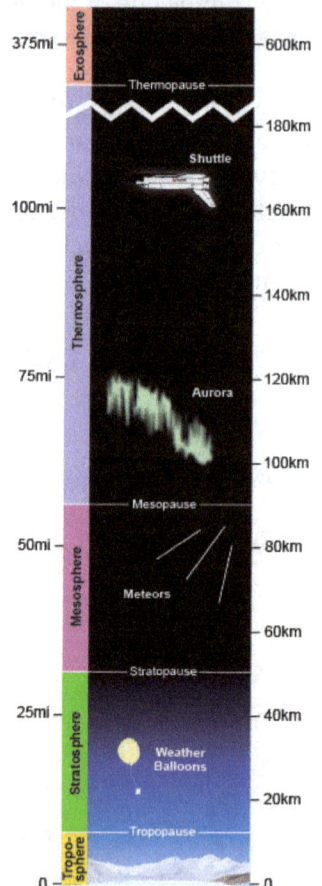

Atmosphere layers: Exosphere, Thermopause, Thermosphere, Shuttle, Aurora, Mesopause, Mesosphere, Meteors, Stratopause, Stratosphere, Weather Balloons, Tropopause, Tropo-sphere

Altitude markers: 375mi / 600km, 180km, 160km, 140km, 100mi / 160km, 120km, 100km, 75mi, 80km, 50mi, 60km, 25mi / 40km, 20km, 0 / 0

The standard model of the sun assumes that the hottest part of the sun is at its core. Sir Arthur Eddington (a good friend of Albert Einstein) was the first to suggest that **nuclear fusion** is happening in the core. Fusion is when small atoms, or small pieces of atoms, are pressed together to form larger atoms. (Sort of like the reverse of the diagram at the bottom of page 4. In fusion, atoms are formed, not broken apart.) Immense heat and pressure inside the sun is believed to drive the process of fusion. Energy is released when fusion occurs, and this energy results in the production of heat and light. If this is true, we would expect to find that the core is the hottest part of the sun, with the layers going outward getting gradually cooler. The corona should be cooler than the inner layers. However, this is not what we find.

This diagram shows the measured temperatures of the outer layers of the sun. (There isn't a way to measure how hot it is deep inside the sun.) The **photosphere** (sometimes called the "surface" of the sun) is about 6,000° C. The layer above that, the lower **chromosphere** is only about 4,000° C. The upper chromosphere then goes up to 10,000° C. The real shocker (as mentioned previously) is that the corona, the layer farthest from the core, is a stunning 1,000,000° C.

This is our first sun puzzle.

Sun puzzle #1: Why is the corona so hot when it is so far away from the core?

Scientists don't have a good answer. Dr. David Brooks of George Mason University made this statement in 2017: "Why the sun's corona is so hot is a long-standing puzzle. It's as if a flame were coming out of an ice cube.

1 000 000 °C *Corona*

10 000 °C *Upper Chromosphere*

4 000 °C *Lower Chromosphere*

6 000 °C *Photosphere*

It doesn't make any sense! Solar astronomers think that the key might lie in the magnetic field, but there are still arguments about the details."

In 2023, a new hypothesis was put forth about this puzzle, using data from the European Space Agency's Solar Orbiter and NASA's Solar Dynamics Observatory. They analyzed the solar corona from multiple vantage points, particularly the arch-like structures called **coronal loops**.

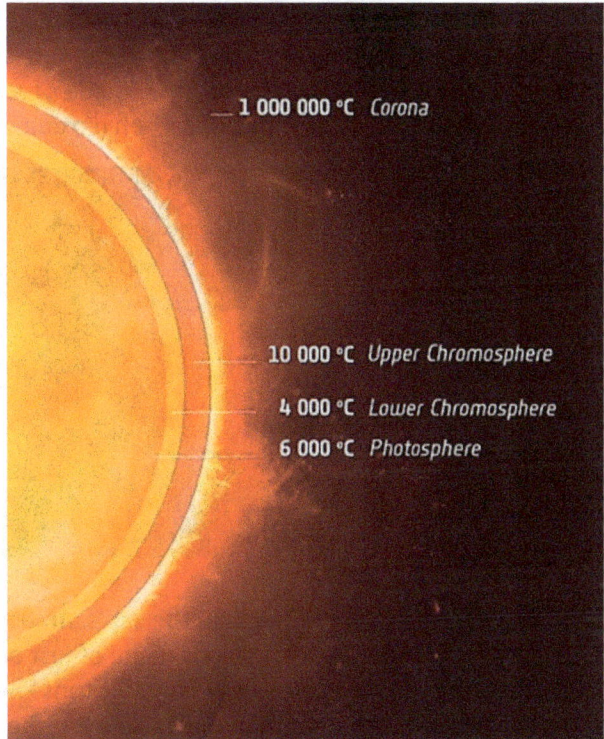

Coronal loops are made of solar plasma, believed to generated by magnetic activity. (But are they?)

The scientists detected a 4-minute "decayless kink oscillation" of a coronal loop. They found that these waves mostly all vibrate in the same direction. This suggests they are "likely associated with long-duration flows in the solar surface," said study co-author Valery Nakariakov, a solar physicist at the University of Warwick in Coventry, England. This discovery provided additional information about magnetic activity in the photosphere, chromosphere and corona, but it did not solve the hot corona mystery. As of 2025, Wikipedia's article titled "Stellar Corona" made this statement: "Many coronal heating theories have been proposed, but two theories have remained as the most likely candidates: wave heating and magnetic reconnection (or nanoflares). Through most of the past 50 years, neither theory has been able to account for the extreme coronal temperatures." This first sun puzzle remains a puzzle to most astronomers.

The second puzzling characteristic of the sun is also found in the corona, the sun's atmosphere. Our atmosphere here on earth is made of mostly nitrogen gas molecules, along with some oxygen and small amounts of carbon dioxide, water vapor, and a few other harmless gases such as argon and neon. We would not expect to find heavier elements, such as iron or nickel, in the air. If tiny molecular amounts of metal get spewed into the air (perhaps by a factory that processes metals) they would eventually drift down to the ground as dust.

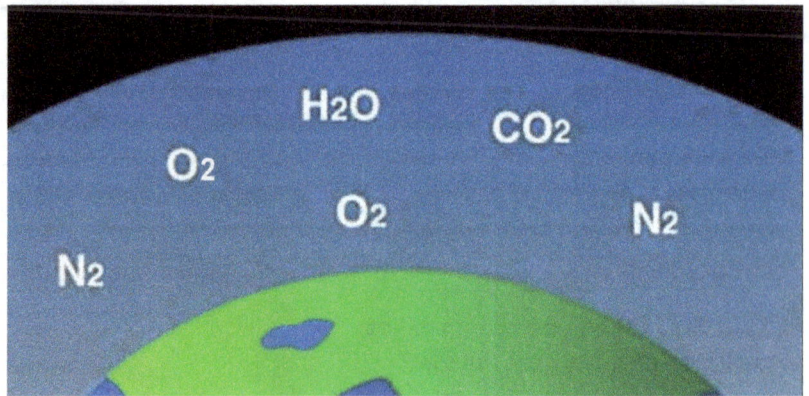

The vast majority of the gas in the sun's atmosphere is hydrogen, which is much lighter than nitrogen or oxygen. Hydrogen is made of just one proton and one electron. There is also a lot of plasma—individual elections and protons—which are also very light. The sun's atmosphere is also much thinner than the earth's—in the range of a million times less dense. (Imagine if the air around you suddenly became a million times thinner!) Thus, we would be even more surprised to find a heavy element like iron in the atmosphere of the sun.

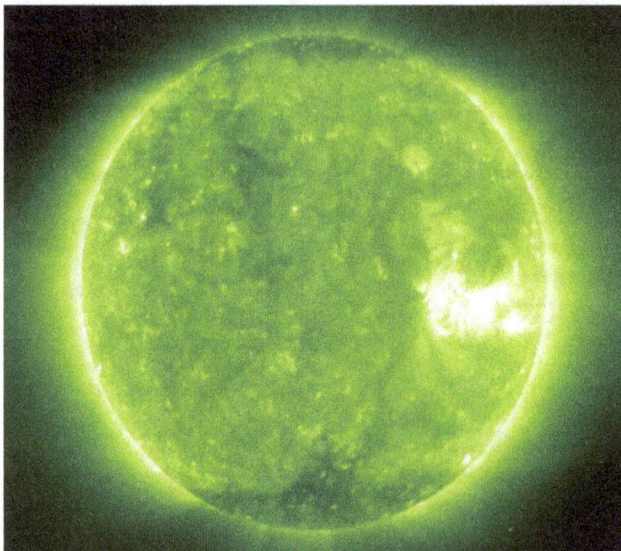

This image of the corona was taken with a special filter designed to detect metallic elements. The green color is due to the presence of iron, which is glowing in the heat. Iron in the sun's atmosphere? One team of scientists concluded that "an unknown mechanism preferentially transports certain elements, such as iron, into the corona, instead of others, giving the corona its own distinctive elemental signature."

Sun puzzle #2: Why is iron (and other heavier elements) found in the sun's corona?

The third puzzle concerns the spots that appear on the surface of the sun. First, some facts about **sunspots**. Sometimes there are many sunspots, and other times there are very few. After many years of counting sunspots, an 11-year patten has emerged. Because of this pattern, rough predictions can be made as to how many spots will appear each year. This is important because more sunspots indicate more solar activity, especially electromagnetic activity that can interfere with satellite and radio communications. Intense surges of solar activity can even affect power grids, causing power outages.

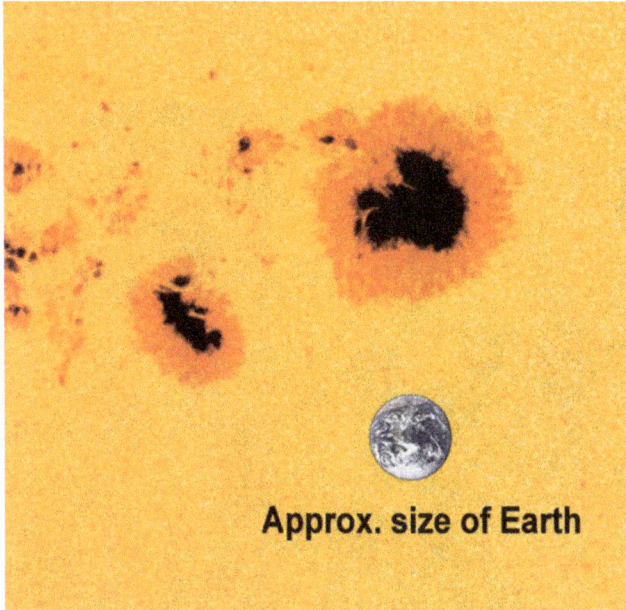

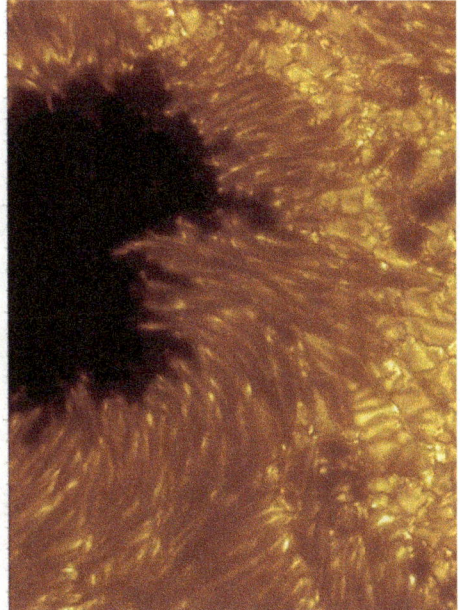

Approx. size of Earth

The image above (left) shows how large sunspots are. They look like tiny dots when you see them on the sun, but in reality they can be even larger than the size of the earth. The image on the right shows the texture of the photosphere around a sunspot.

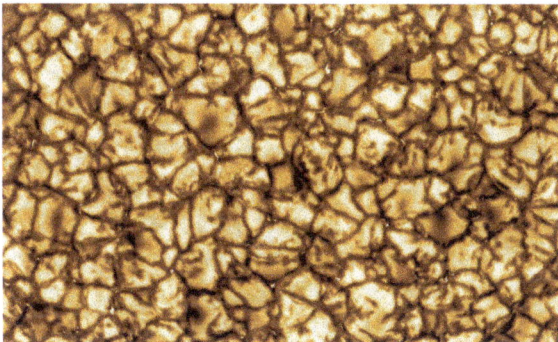

This image shows the texture of the photosphere. The shapes are usually called **convection cells** or **granules**. The cells are constantly changing shape, and constantly appearing and disappearing. It is believed that intense magnetic activity in the sun drives the motion of the cells. When a hole opens up, a sunspot is the result. The spot is black because it is dark inside the hole; no light is coming out.

The mystery associated with sunspots is their temperature. The photosphere around them is about 6,000° C. The temperature of the holes themselves is only 3,000 to 4,000° C. Scientists have been telling us for decades that the hottest part of the sun is its core because that is where nuclear fusion is happening. They say that the immense temperature and pressure in the core is sufficient to smash hydrogen atoms together to make helium atoms. They believe that the heat of fusion flows from the core out to the other layers. If this is true, we should expect that a hole in the photosphere would allow heat from the core to escape. The hole should give us a peek at that immense amount of heat in the core. But this is not what we observe.

Sun puzzle #3: Why are sunspots <u>cooler</u>, not hotter, than the surrounding photosphere?

The fourth solar enigma that we will consider is the **solar wind**. The solar wind isn't actually wind. Wind is the movement of air. Outer space doesn't have any air so it can't have any wind. The solar "wind" is made of charged particles, mostly protons but also containing some electrons, alpha particles (two protons and two neutrons bound together), and a small number of larger positively charged particles. In the illustration below the sun is the tiny yellow dot on the right. The artist has attempted to show that the particles flowing out from the sun are going very fast. They are indeed going very fast—as they leave the sun they are going 250 to 750 kilometers per second. That's *per second*, not per minute or per hour. Try to imagine something covering a distance of hundreds of kilometers in just one second. (That's over a million miles per hour.)

The idea that the sun was spewing particles into space was first proposed by **Richard Carrington** in 1859. He observed a solar flare on the sun. The following day, a geomagnetic storm was observed on earth. Carrington connected the two events and suggested that particles from the sun were affecting the outer atmosphere and the magnetosphere of the earth.

In 1910, British astrophysicist **Arthur Eddington** suggested the existence of what would later be called the solar wind, but the scientific community was not yet ready to accept the idea. The next proposal came from **Kristian Birkeland**. He observed that the northern and southern auroras were almost continuous, although there was variation in their intensity. In 1916, he proposed that the auroras are made of plasma particles coming from the sun.

By the 1930s, scientists had concluded that the temperature of the solar corona must be about a million degrees Celsius because of the way it extended into space, as seen during a solar eclipse. In the 1950s, German astronomers observed that the tails of comets always point away from the sun and they guessed that particles flowing from the sun might be the reason.

In 1957, American physicist **Eugene Parker** coined the term "solar wind." He published his theory about how the solar wind particles escape the sun's gravity in the outer corona and submitted his work to *The Astrophysical Journal* in 1958. The paper was rejected twice before a mathematician came to his rescue and made an appeal to the editor, saying that Parker's work was good, and should be published. In January 1959, the Soviet spacecraft Luna 1 directly observed the solar wind and measured its strength. Three years later, the American spacecraft Mariner 2 confirmed the measurements taken by Luna 1. Since then, there has been no doubt about the existence of the solar wind. In 2018, the Parker solar probe was launched, giving credit to Parker for his work back in the 1950s.

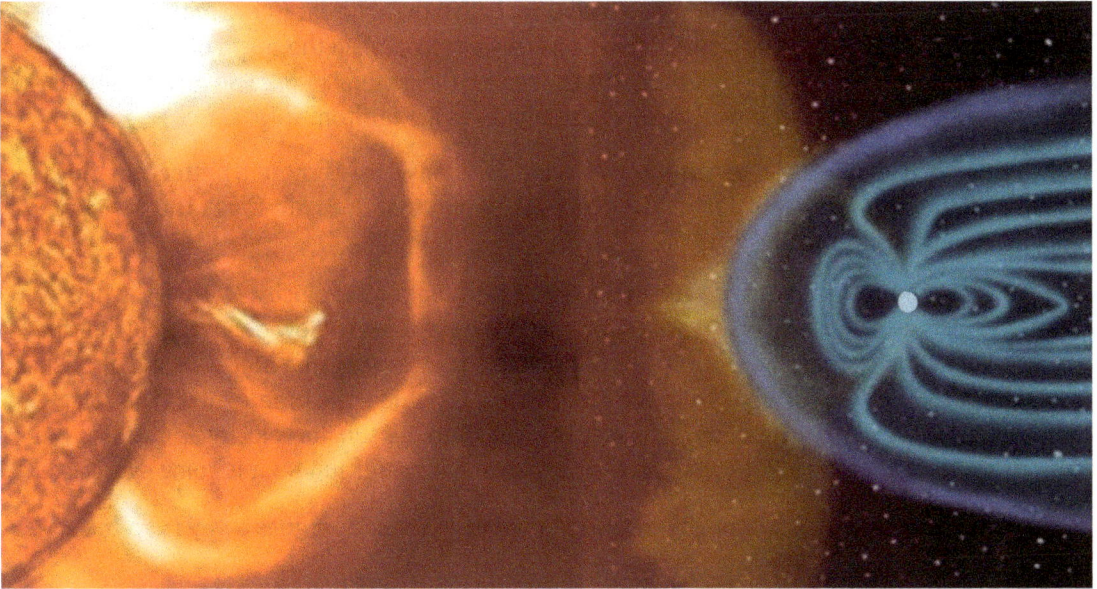

This diagram shows the solar wind (orange) moving outward, beginning to surround the earth's magnetosphere (blue). The earth is the small circle at the center of the blue rings. The particles of the solar wind are moving so fast that if they were to hit the earth, they would cause great harm to living things. Fortunately, the earth is surrounded by a magnetic field that diverts the particles of the solar wind. The blue magnetosphere looks a bit like a wind sock, with its "tail" pointing away from the sun. (This is similar to what happens to the tail of a comet, as the solar wind is always pushing the tails away from the sun.) You can see that the blue lines converge at roughly at the north and south poles of the earth. These are the areas where we see the auroras (Northern and Southern Lights).

The solar wind actually gives us more than one puzzle. First, there are two types of solar wind: the **fast solar wind** and the **slow solar wind**. The fast solar wind speeds away from the sun at approximately 800 km/sec, but it is very thin, containing relatively few particles. The slow solar wind flows at about 400 km/sec and has a large volume of particles. (As a helpful analogy, imagine the fast solar wind as a tiny race car that can go very fast but is too small to carry much of a load. The slow solar wind would be like a dump truck that is heavy and slow but can carry a huge amount of particles. Or perhaps, think of the fast solar wind as a garden hose with the nozzle dialed to "jet," so that a tiny bit of water comes out at great speed. The slow solar wind would be like a very large river with great volume, flowing slowly but steadily.)

According to Wikipedia (which generally represents mainstream thought), the fast solar wind originates from "coronal holes" which are funnel-like regions of open field lines in the sun's magnetic field (located 20,000 km above the photosphere). The plasma particles come from small magnetic fields created by those convection cells we saw on the previous page. The plasma is released into the "funnels" whenever the magnetic lines reconnect. The slow solar wind appears to originate from around the sun's equator (equivalent to earth's tropical climate areas) where "coronal streamers are produced by magnetic flux open to the heliosphere draping over closed magnetic loops." Other websites often use the term "open magnetic field lines" in their explanations. Electrical engineers should be quick to point out that these astronomers feel free to ignore basic rules of electromagnetic theory when proposing theories about the sun. There are no "open field lines" in electromagnetism (according to Maxwell's equations, which no one disputes). Is there a reason why the sun is exempt from basic laws of physics?

The standard model of the sun (based on the idea that nuclear fusion in the core is the ultimate origin of all of the sun's characteristics) struggles to explain the origin of these two different types of solar winds, even though websites for the general public will make it sound like their theories are settled science. However, these puzzles are far from solved.

Sun Puzzle #4: What is the origin of the two types of solar wind? Is there a better and much easier explanation for these solar winds— an explanation that does not break the basic laws of how electromagnetism works?

A second puzzle about the solar wind is the behavior of its velocity. Normally, when anything flows away from a certain area, you expect the particles to slow down the further they go. As a very simple example, if you sit in front of a fan, you feel a strong breeze as the air rushes past your face. If you sit far away from the fan, you might not feel the breeze at all. The solar wind doesn't act like this. It speeds up as it goes away from the sun. Its velocity as it passes the earth is much faster than when it passed Mercury. When it passes Jupiter it is going even faster.

Sun Puzzle #5: Why does the velocity of the solar wind increase as it goes away from the sun?

A third puzzle about the solar wind:

Sun Puzzle #6: In 1999, the solar wind almost stopped for several days. How could this happen?

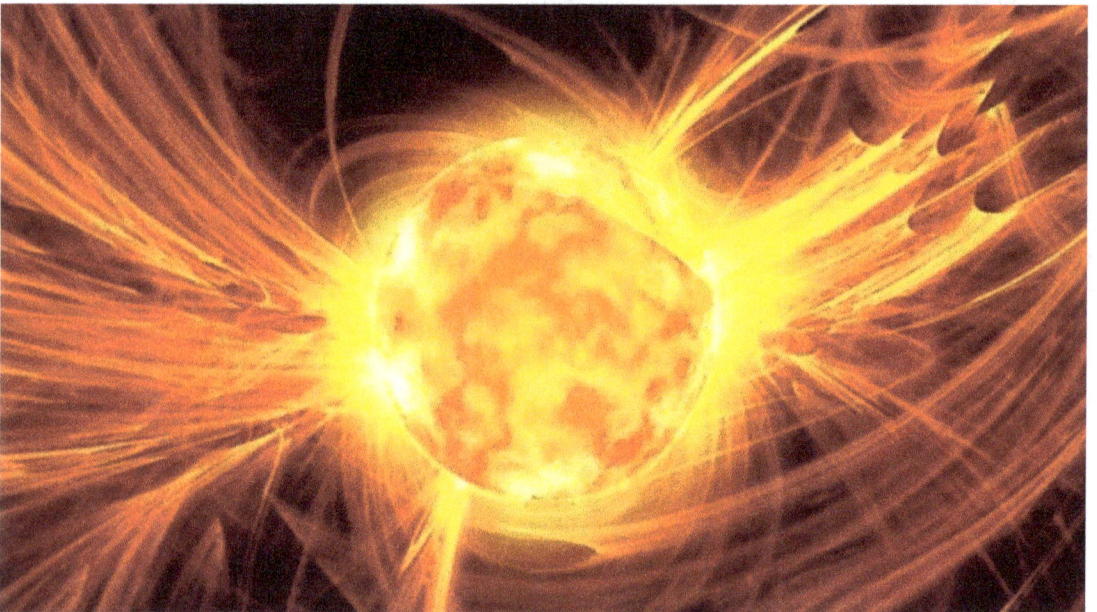

The answers to these puzzles about the sun have been sitting in plain view for decades. To see how these answers came about, we need to highlight the work of a few key scientists. About 100 years ago, Kristian Birkeland (introduced on page 8) realized that the auroras were connected to a system of electrical currents that came from the sun and flowed through the earth's upper atmosphere at the poles. His ideas were not accepted by the mainstream scientific community, and were ridiculed by many. No one believed that electrical currents could cross the vacuum of empty space. Because his ideas were not given serious consideration (until 1967 when a probe confirmed he was right), mainstream science continued on with weak theories.

Kristian Birkeland (1867-1917)

Birkeland tested his ideas as best he could in his lab, using equipment that was available at the turn of the 20th century. His "terrella" experiment is shown above. The ball in the center was used to represent the earth. Air was removed from the clear box, and then plasma was generated inside. He was able to make some basic observations about how plasma works, enough to confirm his suspicions about the nature of the auroras.

Birkeland realized the importance of plasma in astronomical science, and may have been the first scientist to propose that outer space is filled with plasma. In 1913, he said, "The whole of space is filled with electrons and flying electric ions of all kinds." (An "ion" is any particle that has an electrical charge, either positive or negative, but in plasma science the word "ion" tends to be used more often to describe positive particles, and electrons are just called electrons.) The importance of plasma in space will turn out to be critical to solving our sun puzzles.

Hannes Alfvén picked up where Birkeland left off, though their careers did not overlap. Alfvén was still a child when Birkeland died. Alfvén shared Birkeland's interest in auroras, perhaps because both of them lived in northern countries where the auroras were often visible. Alfvén followed up on Birkeland's research on magnetic storms, the earth's magnetosphere, the possible existence of trapped solar particles in belts that encircle the earth (which were eventually named Van Allen belts), and the implications of great amounts of plasma in the Milky Way galaxy. Alfvén's research was first-rate, and he received the Nobel Prize in physics in 1970 for his work in a branch of science called magnetohydrodynamics.

Hannes Alfvén (1908-1995)

Despite his excellent work, and the Nobel Prize that established him as a leading pioneer in the field of plasma physics, Alfvén had outspoken critics who were unwilling to consider his ideas and often prevented his papers from being published in major journals. British scientist **Sydney Chapman** was one of the harshest critics, accusing Alfvén of espousing "unorthodox opinions." He had also been very critical of Birkeland's work. Chapman may eventually go down in history as someone who prevented the advancement of science. Both Birkeland and Alfven had discovered important clues about how plasma works— clues that would eventually lead to a highly successful model of how the sun, stars, and galaxies work.

Another scientist who should be mentioned at this point is American physicist and chemist ***Irving Langmuir*** (1881-1957). Langmuir won many awards and medals in his lifetime, including the Nobel Prize in chemistry in 1932. He made many practical discoveries, such as the use of inert gases inside light bulbs, the curling of the tungsten wire to make the bulbs last longer, and, later, the invention of the hydrogen welding torch. His most controversial invention was the "seeding" of clouds with silver iodide to create rain.

Langmuir and associates in a G.E. lab

His work with electric bulbs during his life-long career at General Electric in New York led him to begin investigating plasma, and it was he who coined the word "plasma" because it reminded him, oddly enough, of blood plasma (the watery portion of our blood). He invented a probe that could be put into a plasma to take its temperature.

Langmuir's contribution to the answers to our sun puzzles lies in his discovery of a phenomenon that he called "double sheaths." He found that plasmas can maintain two distinct, very thin layers—one positive and one negative. This was a surprising finding because it had been assumed that positive and negative charges <u>always</u> attract, and therefore, <u>always</u> come together to "cancel each other out." Scientists would have agreed that it was impossible to have areas of opposite charge next to each other but staying separated. Yet there it was—this was happening inside Langmuir's plasma tubes. Later, these layers would become known as "double layers." The opposite charges could stay separated because they were moving in opposite directions. The charges just "stayed in their lanes." Years later, Alfvén would realize that double layers could exist in space, as well as in labs. He speculated that double layers were a key to explaining how the auroras worked. (And he was right.)

The work of these three scientists laid the experimental and theoretical foundation for what should have become the mainstream explanation for how the sun works. These scientists were well-respected researchers, not crazy crack pots. However, history took an unfortunate turn at this point. The era of atomic science (1930s to 1950s) came into the story, with its intense focus on the processes of fission (splitting atoms) and fusion (forcing smaller atoms together). Science

acquired a "one track mind" during this time, applying its nuclear energy bias to the sun. Something as large and powerful as the sun must be using either fission or fusion to release energy. They didn't consider other options. They concluded that fusion had to be the answer, and schools and textbooks have been teaching the fusion theory ever since. Students learn that the core of the sun is a nuclear fusion reactor where hydrogen atoms are smashed together to make helium atoms, releasing energy in the process. This is taught as fact, even though we have no way to observe what is going on inside the sun. Anyone who questions the nuclear fusion theory is either ignored or ridiculed.

During the 1960s, an American engineer named **Ralph Juergens** began reading about the research that Birkeland and Langmuir had done, and that Alfvén was currently doing. He also took note of an interesting statement at the end of a 1958 paper titled "The Science of High Explosives" by Melvin Cook. Cook suggested that the sun's radiant energy might not be solely the result of nuclear fusion, but that electrical properties could also be involved. By 1972, Juergens had digested all of their ideas about plasma and electricity, and wrote a paper in which he introduced a theory that he called "The Electric Sun." He wrote: "The known characteristics of the interplanetary medium suggest not only that the sun and the planets are electrically charged, but that the sun itself is the focus of a cosmic electric discharge—the probable source of all its radiant energy."

Juergens' paper was not enthusiastically received by the scientific community; it was either ignored or criticized. Juergens was accused of being imaginative and incompetent. Physicists and astronomers were offended that a lowly civil engineer would dare to "drive out of his lane" and propose an astronomical theory. This attitude continues to this day. If you search Wikipedia (the repository of all things mainstream) for "Electric Sun" you get information on a rock and roll band by that name. If you search for "Ralph Juergens" you get no result at all, despite the fact that they have articles about the authors of much more obscure theories, as well as thousands of articles on fictional characters.

Other scientists and engineers have continued to refine Juergens' original electric sun model but before we get to them, let's stop and learn some more basics about plasma, since plasma science will be the key to explaining the electrical properties of the sun and the answers to the sun puzzles.

You will remember that a plasma consists of electrically charged particles: negative electrons, positive protons, and small clumps of protons and neutrons that have an overall positive charge. Plasmas that we are familiar with are shown below.

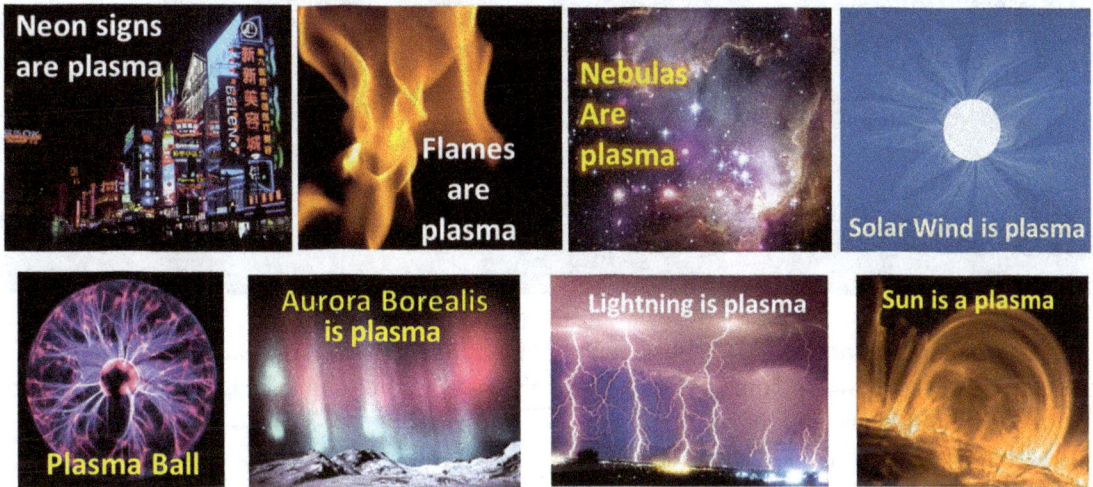

Neon signs are plasma

Flames are plasma

Nebulas Are plasma

Solar Wind is plasma

Plasma Ball

Aurora Borealis is plasma

Lightning is plasma

Sun is a plasma

Plasmas can exist in one of three modes: *dark mode, glow mode, and arc mode*. Dark mode plasma can't be seen. When you look up at the night sky, you are looking at dark mode plasma. The solar system is filled with dark plasma from the solar wind. The earth's atmosphere has a layer called the ionosphere, which is made of ions (charged particles) that form a thin plasma. Glow mode plasma can be seen inside fluorescent and neon lights and in the auroras. If a glow mode plasma becomes even more energized, it can go into arc mode. Lightning is made of arc mode plasma. Even tiny sparks of static electricity are made of arc mode plasma. All stars, including the sun, have arc mode plasma on their surface (photosphere). The plasma sometimes forms loops called coronal mass ejections (CMEs), as shown in the lower right, above.

The existence of plasma in the solar system and in outer space isn't controversial. Nebulas are known to be plasma formations, though they often vary in visibility. We see beautiful color photos of nebulas and assume that this is what you would see if you looked through a telescope. However, images are often made with cameras that detect infrared radiation or even x-rays or gamma rays. Both images below are of the same nebula: the Crab Nebula.

As our telescopes get better, we are able to see farther into space and in greater detail. Plasma filaments have been discovered that are several light years in length.

Kristian Birkeland and Hannes Alfvén predicted that we would probably find plasma filaments in space. Birkeland died before he got to see any photographic images of these plasma filaments, but Alfvén lived into the 1990s, and had the satisfaction of knowing that many of his predictions have been confirmed. There are, indeed, plasma filaments in outer space. In fact, it has recently been discovered that the entire universe has a filamentary structure. The image on the right shows a "map" of the filaments in the universe. This shocked most astronomers. But how do filaments in space help us to solve our sun puzzles?

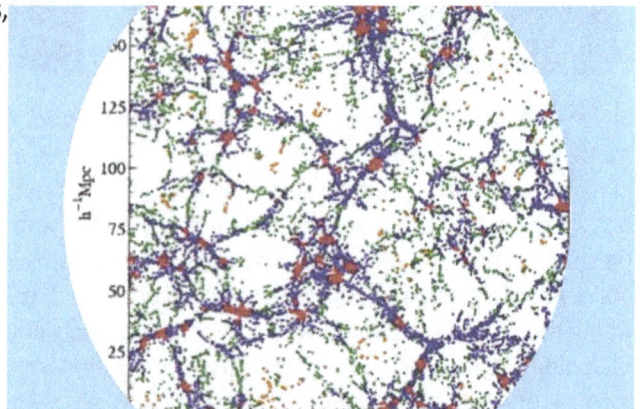

These plasma filaments in space are often carrying a massive amount of electric current. This is because any time you have charged particles in motion, you automatically have a current.

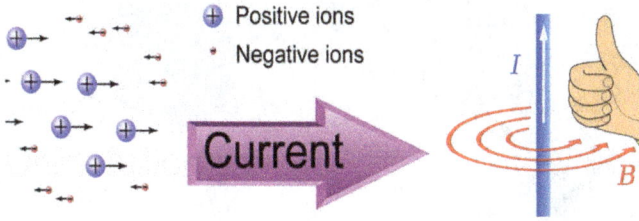

Positive ions

Negative ions

Currents (I) always have circling magnetic fields (B) constraining them

Electrical currents always have a magnetic field encircling them and constraining them. Electricity and magnetism are inseparable. If you have one, you have the other. Both of them have a direction, as shown by the hand in this diagram. If you point the thumb of your right hand in the direction in which the charged particles are moving, the fingers of your hand will show you the direction of the magnetic field.

Because electricity and magnetism are linked together, we often refer to this phenomenon as *electromagnetism*. Electromagnetism is one of the four fundamental forces (that we know of) in the universe. The others are *gravity*, *the strong force* (that keeps an atom's nucleus bound together) and *the weak force* (which is involved in the complicated process of a neutron turning into a proton and an electron). Gravity is by far the weakest of these forces, and the strong force is by far the strongest, but electromagnetism is much stronger than gravity. In fact, electromagnetic forces in space have been measured to be up to 1,000,000,000,000,000,000,000,000,000,000,000,000,000 (10^{39}) times stronger than gravity.

This diagram shows a plasma filament observed in a lab experiment. The gray color represent the electric current and the black circles represent the encircling magnetic field. The tight spot in the middle represents some kind in instability—a region where turbulence occurs due to changes in the characteristics of the plasma (for example, temperature, density, electric fields, or magnetic fields). Instabilities can lead to the plasma forming a "pinch" (known as a *Z-pinch,* or *Bennett pinch).* The pinches have been observed to turn into a bright plasma ball (a plasmoid).

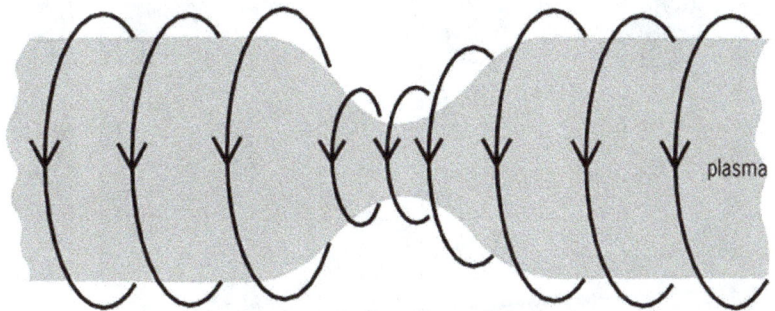

The image on the right show the Z-pinch research machine at Sandia Nat'l Lab in New Mexico. It requires a lot of energy.

If a plasma filament in space is pinched, the same things happens—a bright plasma ball forms at the area of the pinch. However, these cosmic filaments are so large that the plasma ball is huge. We call these plasma balls "stars." The nebula on the left is called the Ant Nebula. You can clearly see the star at the center. The one on the right is called the Butterfly Nebula.

Many nebulas have a symmetric shape. (Why?) Some have a visible plasma ball in the center, but in others the central star (or quasar) has dimmed in the years since it first lit up.

The Heart and Soul Nebula

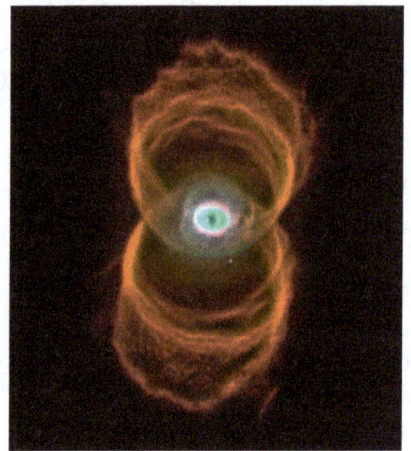

The Hourglass Nebula

Galaxies are also plasma structures. The image on the right shows an extremely distant galaxy (NGC 4038) that was recently discovered using the newest space telescopes. Astronomers were surprised to see this much structure in a galaxy they believe to be relatively young. However, plasma scientists were not surprised at all, because they understand that electromagnetic forces (10^{39} stronger than gravity) probably played a significant role in both star and galaxy formation.

NGC 4038

American physicist **Anthony Peratt** (still living) was a graduate student under Hannes Alfvén in the late 1960s. After receiving a PhD in electrical engineering in 1971, he worked on plasma physics at Lawrence Livermore National Lab, then at Los Alamos National Lab. During this time, one of the experiments he conducted was to see what would happen if two plasma filaments were brought close together. In the first picture on the top left, you are looking at a cross section of two filaments. Imagine the filaments being perpendicular to this page, coming out of the page straight at you. As these filaments get closer, you can see them changing shape and starting to interact. In the bottom left image, you can see that they have begun to swirl together. The last of the eight images shows their final shape... looking very familiar.

These images show galaxies NGC 1365 (left), M81 (center) and the Milky Way (right). Not all galaxies have this shape, but many do. Perrat was able to generate every galaxy shape by varying the way the filaments interacted.

Perrat also noticed another very important event in the formation of his "mini-galaxies." Tiny pinpoints of arc mode plasma (tiny plasma balls) appeared on the spiral arms of these galaxy structures. If these tiny plasma balls were equivalent to stars on large galaxies, we should see stars mainly on their spiral arms. A close look at a real spiral galaxy reveals that they do have this same structure. One astronomer was so impressed by this that he described the stars as looking like they were "beads on a string."

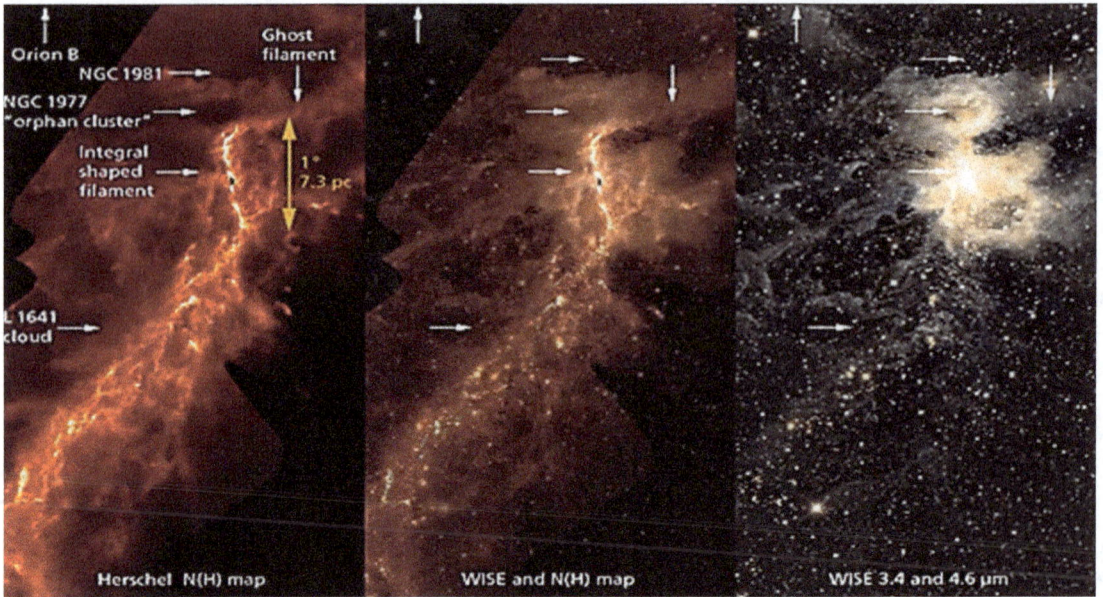

Herschel N(H) map | WISE and N(H) map | WISE 3.4 and 4.6 μm

Here we see a filament visible in the constellation Orion. The three photos are of exactly the same area, but were taken with different types of light. The star "beads" along the filament are best seen in the red and brown photos. Many of the stars in the photo on the right are not part of the filament; they are light years in front of, or behind, the filament.

Plasma physics has a workable hypothesis for how and why stars form from plasma filaments. (These will eventually lead us to the answers to the sun puzzles.) But first, what about standard astronomy—how do they explain star formation?

The standard model for star formation is called the Nebular Hypothesis. It assumes that as a result of the Big Bang, massive amounts of gas and dust were floating around. As part of a random process, some gas and dust began collecting in one spot more than in others. Since everything has gravity, these small particles also had gravity. The gravity of the collected particles began attracting even more particles. Eventually, there were enough particles gathered together that the gravity became strong enough to make the clump collapse, forming a star.

This sounds very simple and easy but there are problems. We know how particles behave when they get closer together: they begin hitting each other more frequently. As particles

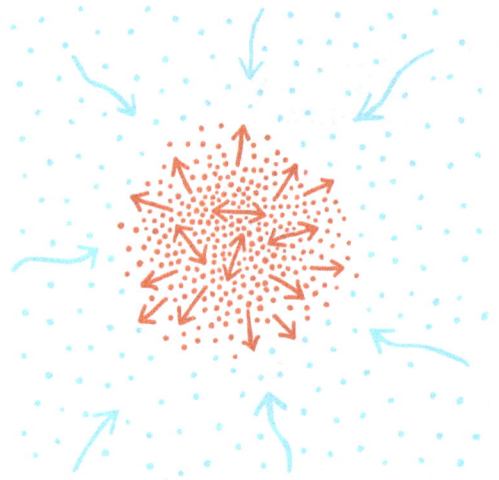

gravity in blue, heat/kinetic motion in red

collide and bounce off each other, the increased kinetic motion drives them apart. The molecules in question are not being confined in a container. They are free to travel away from each other; they will not try to stick together. There is a "tug of war," so to speak, between gravity (in blue) and the kinetic motion (in red) of the particles. Gravity is a very weak force, and there is little gravitational attraction between the particles. How do you grow larger when your particles are always flying apart?

Astronomers do recognize this problem. They call it "the accretion problem" or the "fragmentation barrier." They assume that there are lots of collisions going on all the time in the universe and that collisions might destroy any small clumps that had begun to form. The clump has to be big enough to sustain hits. Needless to say, there's more than one theory about how proto-stars overcome this problem. Special conditions and dark matter are often invoked. Time is also a key—statistics dictate that billions of years are needed in order for everything to finally go right and result in star formation.

The plasma explanation of how stars form invokes a very strong force, the electromagnetic force, and doesn't need any special conditions, or dark matter, or even eons of time. As further evidence that the plasma explanation should be seriously considered, when plasmoids form in the lab, streams of charged particles ("polar jets") an be seen coming out of either end.

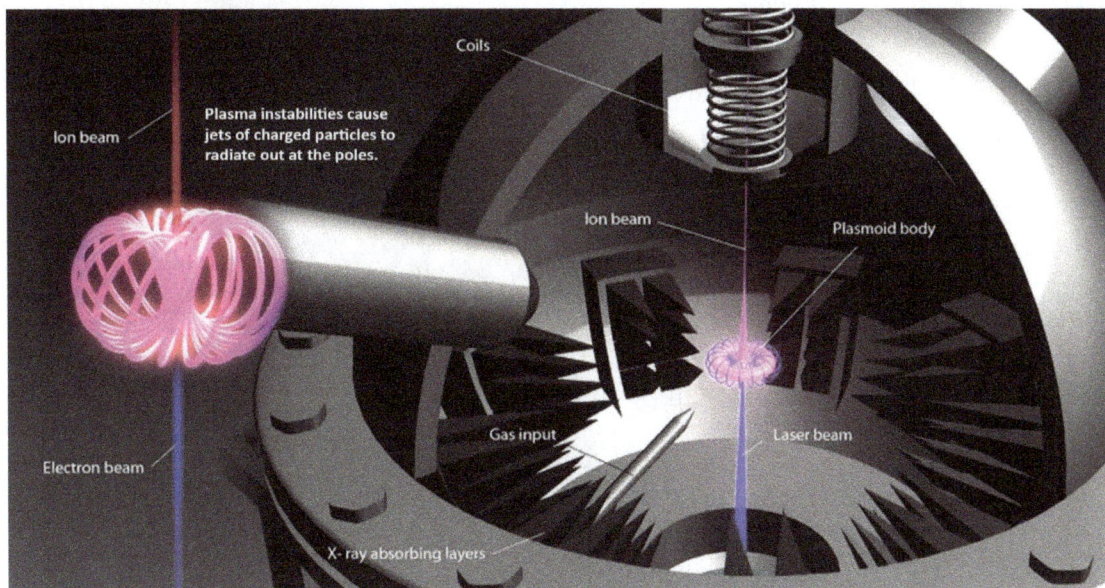

If the plasma theory is correct, we should expect to find this polar jet structure somewhere out there in space. We do find it at extreme distances in special stars called *quasars*.

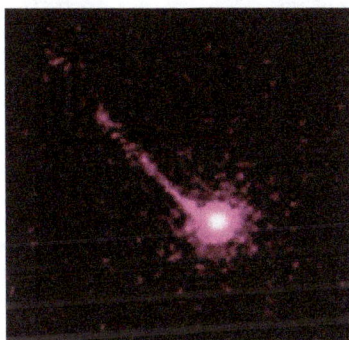

Quasar PKS 1127 145

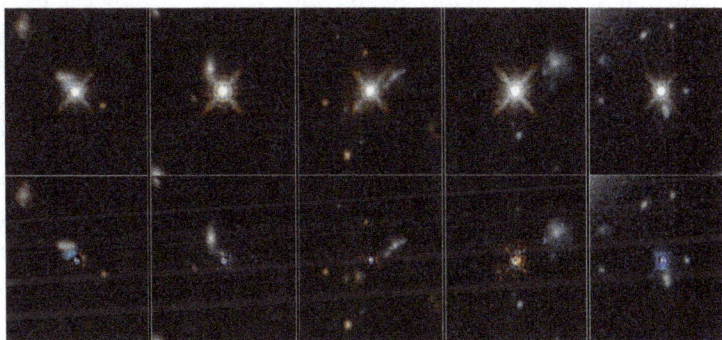

Hubble photos of quasars (NASA image)

The image on the right is an artist's painting of what a quasar would look like if you could see it up close instead of in a telescope. This illustration helps us to see the polar jets. Astronomers know that the jets are there, even though they can be hard to see in actual telescopic photos.

If we assume that the sun is a plasma ball, and not a nuclear furnace with fusion happening in its core, our sun puzzles are easily answered. Let's tackle puzzle #3 first.

Sun puzzle #3: Why are sunspots ("holes" in the photosphere) cooler?

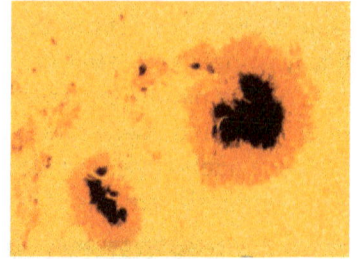

If the sun is a plasma ball, and there isn't any fusion happening in the core, there is no need to be surprised at the lower temperature of sunspots. It is possible that the sun is a uniformly dense sphere of gas, possibly with a slightly denser core. No one knows what is at the center of the sun because we can't peer down into the sunspots. However. there is more evidence on the side of "no fusion" than "fusion."

If we address puzzle #4 next, the answers to the other puzzles will fall into place.

Sun Puzzle #4: What is the origin of the two types of solar wind?

To answer this, we need to learn more about the structure of the photosphere. As previously mentioned, in the standard model the odd geometric shapes that cover the photosphere are called "convection cells" or "granules." The standard model teaches that each cell convects, similar to what boiling water does in a pot on the stove. The center of each cell is believed to be the hottest part, where hot plasma comes up from below. The plasma flows out to the edges of the cell, becoming cooler and therefore sinking back down. (The standard model says that the surface of the sun is made of plasma, but without any electrical explanation.)

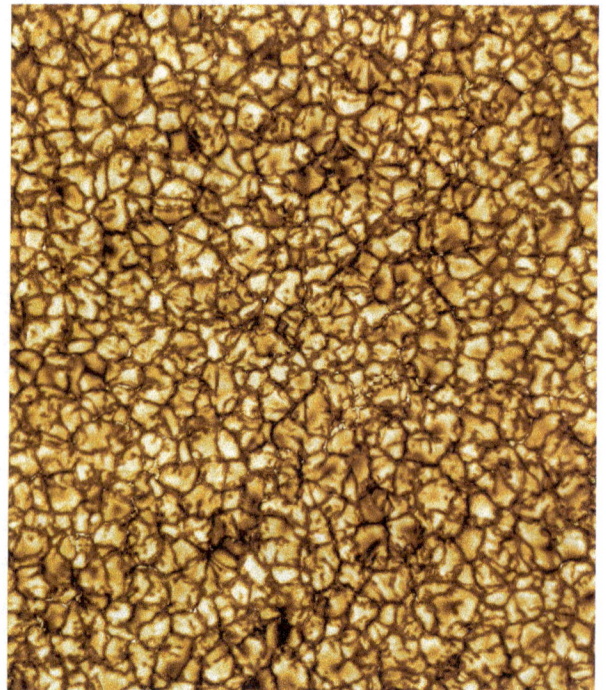

"Convection cells" ("granules") on the surface of the sun Each cell is over 1,000 km wide!

By NSO/AURA/NSF - Daniel K. Inouye Solar TelescopeHigh res URL (archived link)License details (Archived URL)Image URL (Archived link)Description (Archived link)Date taken (Archived link), CC BY 4.0, https://commons.wikimedia.org/w/index.php?curid=86429003

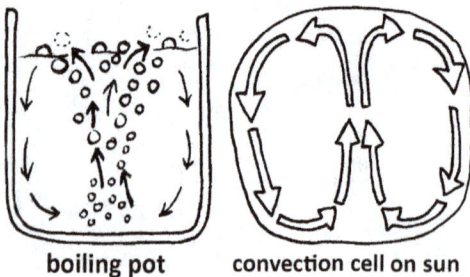

boiling pot   **convection cell on sun**

These cells are constantly forming, changing, disappearing and reforming, the way you see bubbles come and go in a pot, more slowly than pot bubbles, but still visibly changing as you watch.

Is there anything wrong with the convection cell explanation? According to Ralph Juergens, "Photospheric granulation is explainable in terms of convection only if we disregard what we know about convection." Engineers use a term called the "Reynolds Number" to describe the characteristics of things that flow. This number takes into account temperature, viscosity, density, and other measurements. When these things are measured for the sun and then used to calculate its Reynolds Number, it turns out the sun's Reynolds Number is 100 billion times too large for convection to work. This doesn't seem to bother astronomers who like the standard model; they go right on with it despite this problem.

Plasma scientists prefer to call these granules **anode tufts**. The anode tufts are where positive charges (mostly protons) are escaping while negative electrons are entering. There isn't any convection, but particles are definitely moving around in the tufts, causing them to change shape, grow, shrink, appear, and disappear.

The electric sun model suggests that the interior of the sun is filled with positive ions (primarily protons). They are not being fused together (no nuclear fusion) but they are moving around very quickly. Since positive charges repel each other, these ions are "looking to escape." If there is a way to burst out, they will try to do so. The place where the positive ions are leaking out is around the edges of the anode tufts. However, the electrical situation in the photosphere is such that escape isn't very easy. The voltage situation on the surface of the photosphere is such that the positive ions will have to climb a voltage "hill."

Time out: let's discuss **voltage**. Voltage is the difference in electric potential between two points. Sometimes it is referred to as electrical "pressure." Voltage is defined so that negatively charged objects are pulled towards higher voltages, while positively charged objects are pulled towards lower voltages. A common analogy for voltage is a water hose that has an adjustable nozzle. When the nozzle is open wide, a lot of water comes pouring out, but it does so fairly gently. If the nozzle is narrowed, the pressure of the water will greatly increase so that the stream of water is very intense and can blast dirt off a sidewalk. The water coming out is like electrical current (flow of electrons). The current can be large but with low voltage. Or, under pressure, the current can be small with very high voltage.

Another common analogy is a sledding hill. When you are at the top of the hill, there is a large difference between your position there and your previous position at the bottom of the hill before you made the climb up. The higher the hill, the greater the difference in potential energy. You'll get a faster and longer ride from a very high hill because you will have a great amount of potential energy. You turn that potential energy into kinetic energy (motion) as you slide down the hill. You might measure your "voltage" by the height of the hill.

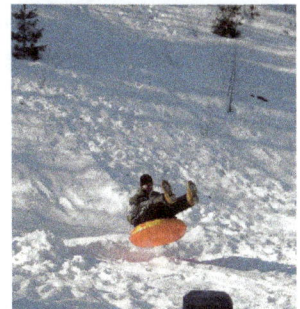

These analogies aren't perfect, of course. Electrical voltage can't be totally explained by simple non-electrical analogies. But it is helpful to imagine protons trying to climb a voltage hill— that hill is the side of one of the anode tufts. A few protons manage to make it to the top. Once they are on top, they can stay there for a while, but eventually they might slide off the far side of the hill, giving them a very fast ride away from the sun. This is the source of the **fast solar wind**. This process can be represented by the graph shown below.

This might look like a regular hill, but it's a bit more abstract than that. The downward slope that occurs at point "c" is a voltage drop, not a literal hill. The protons don't actually "fall" down the hill, they accelerate away from the sun. They continue to feel an electrical "push" as they travel away from the sun and into the solar system.

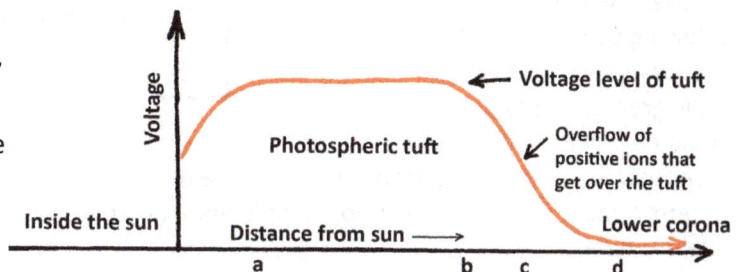

Voltage level of tuft

Photospheric tuft

Overflow of positive ions that get over the tuft

Inside the sun

Distance from sun ⟶

Lower corona

a   b c   d

The fact that the solar wind speeds up as it get farther from the sun is not a characteristic predicted by the standard model. **(This was our Sun Puzzle #5.)** If the particles of the solar wind are being flung out from the corona, similar to the way particles are flung away from an explosion, you would expect them to slow down as they got farther away. The electrical explanation makes sense of the surprising acceleration. The sun has an overall positive charge, and so do all the positive ions streaming away from it. This repelling force extends for millions of miles. Additionally, there is a possibility that the edges of the solar system are negatively charged, so the positive ions might be actively attracted by that negativity. Eventually, the fast solar wind slows down, but not until it reaches the most distant areas of the solar system.

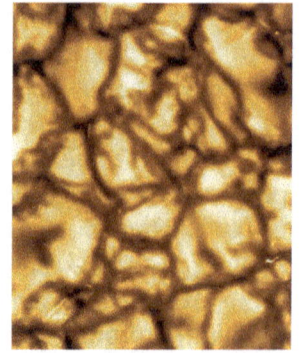

The fast solar wind comes from the anode tufts.

To explain the **slow solar wind**, we need to take another look at the graph we saw on the previous page, and compare it to a water dam. The reservoir of water behind the dam is equivalent to the reservoir of positive ions inside the sun. The thin layer of water at the very top of the dam is like the highest voltage level at the top of the anode tuft. The water spilling out over the dam is like the protons that are following the drop in voltage from the photosphere to the lower corona.

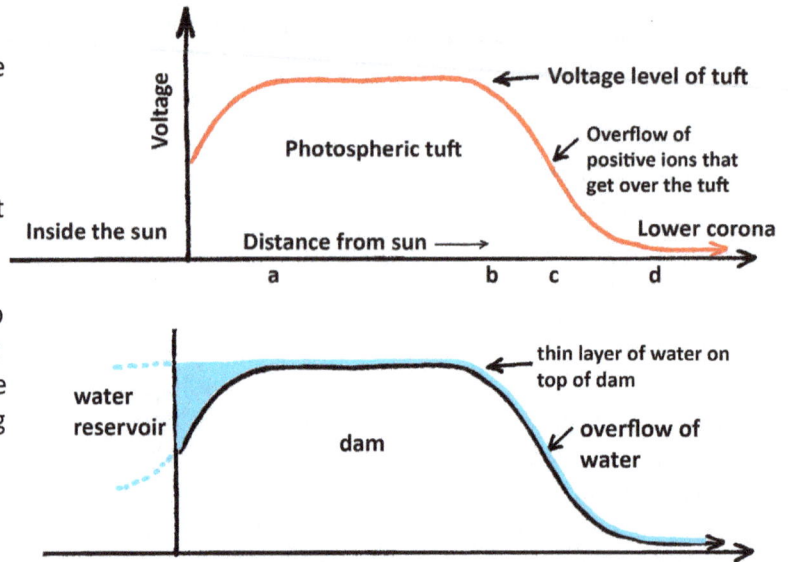

In the water dam analogy, what controls the speed (force) of the overflowing water as it comes racing down the hill? It isn't the amount of water that is flowing, it is the height of the dam. If the lake is shallow and the dam is only 10 meters high, the water will only be falling a distance of 10 meters. However, if the dam is 100 meters high, the water will be falling a distance of 100 meters and will be able to accelerate quite a bit during that long fall. The amount of water does not have to increase for the water to have more force (more voltage).

Now imagine that someone punched a large hole in the dam—a hole that let out a considerably larger volume of water than what is flowing over the top. The water would come gushing out in great quantity but would be flowing with less force than the water falling from the top of the dam. More water, but lower velocity. (In electrical terms, more current but less voltage.) Is there anything on the sun that looks or acts like a hole in a dam?

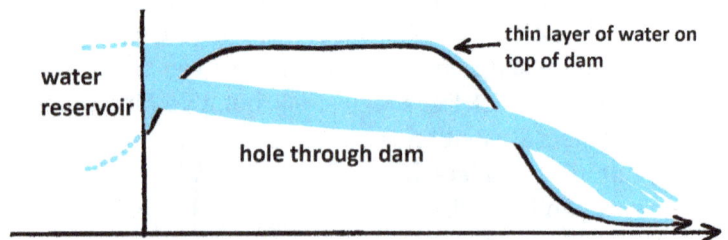

Sunspots are holes in the photosphere. The slow solar wind seems to be coming from these holes. It has a lower velocity but a higher density than the fast solar wind. Using the water analogy, more water is coming out, but at a slower speed. Sunspots tend to occur in the equatorial regions of the sun, between 20 degrees north and south of the sun's equator. Occasionally, a sunspot will drift out of this zone, but as a general rule, this is where we find them. Is it a coincidence that the slow solar wind seems to be coming from this part of the sun?

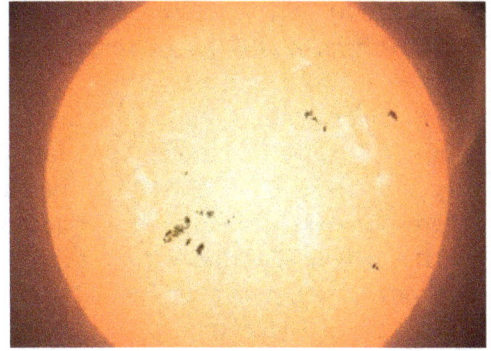
Sunspots come and go over days and weeks.

Sun Puzzle #6: On May 10-12, 1999, the solar wind almost completely stopped. It went from its normal 5 to 10 protons per cubic centimeter down to only .2 protons per cubic centimeter. If the standard model is correct, how could this happen?

If the sun is being powered by a nuclear fusion furnace at its core, and the solar wind comes from particles being spewed out as an end result of this process, how could fusion stop for several days, then restart again? The fusion model has no good explanation for this strange phenomenon.

In the electric sun model, the water dam analogy can help us come up with an answer. In this cross section of a dam, you can see a flood gate. This gate can be adjusted up or down to control the amount of water that is allowed to pour out over the dam. It might be possible to raise the flood gate enough so that no water escapes. In electrical terms, if the voltage goes up in the anode tufts, fewer protons will be able to make it up to the top and therefore fewer will "ride" down the voltage drop.

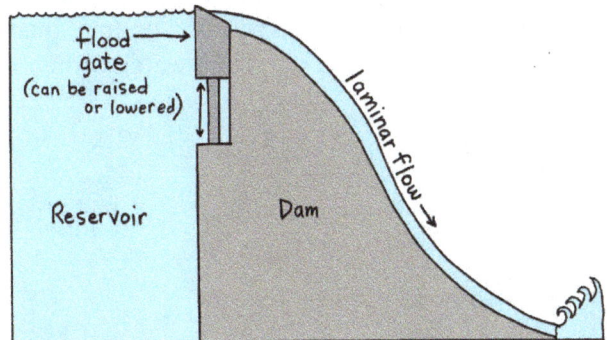

What could make the voltage go up in the tufts? While we will never know for sure what happened, we do know that the sun experiences many fluctuations. It gets brighter and dimmer, and gets bigger and smaller (though these changes are too small for us to detect without scientific instruments). The entire solar system is moving through space at a speed of about 720,000 kilometer per hour. Therefore, the sun is constantly moving into "new" space that it has never been in before. Outer space is not the same everywhere. It is full of charged particles and electrical currents that vary from place to place. Perhaps on those days in 1999 it passed through an area whose electrical properties affected the voltage of the tufts. While we can't be dogmatic about this answer, it is consistent with the electric sun model and makes sense. The standard model does not have a reasonable explanation for how the solar wind could stop.

Birkeland (electric) currents in space can be many light years in length. A light year is the distance that light travels in a year. Light travels at about 300,000 kilometers per second.

Sun Puzzle #1: Why is the corona so hot?
(Bonus Puzzle: Why is the chromosphere so cool?)

This diagram reminds us of the puzzling temperature zones in the sun. We've just discussed the photosphere quite a bit. The photosphere is often referred to as the surface of the sun. It is an area of arc mode plasma (page 11 for review). The temperature minimum of the sun is at the bottom of the chromosphere. Then the temperature begins to rise very quickly, as we get into the atmosphere of the sun, the corona. Although these observations elicit comments such as "bizarre," "mysterious," or "surprising," from standard astronomers, scientists who have a good understanding of electricity and plasma are not in the least bit surprised.

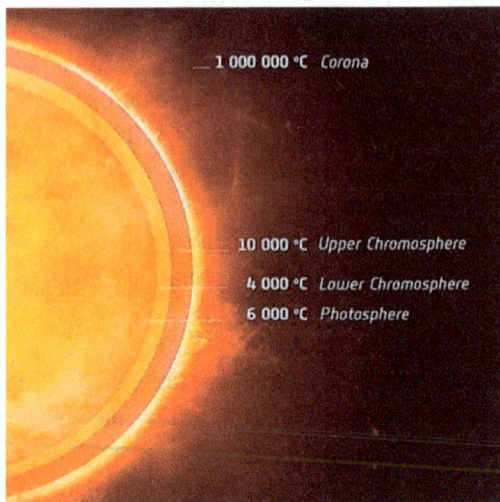

1 000 000 °C *Corona*

10 000 °C *Upper Chromosphere*

4 000 °C *Lower Chromosphere*

6 000 °C *Photosphere*

Once again, the water dam analogy will be very useful. If the water falling over the dam hits another body of water at the bottom, there will be a powerful churning action as the high speed water hits the stationary water. Or think of a waterfall, with powerful rapids at the bottom. In the sun, the positive ions are going very fast down the voltage drop, away from the photosphere. When the fast-moving ions hit the relatively stable corona, they cause the ions in the corona to "churn." The definition of temperature is the measure of how fast atoms or ions are moving, so this massive churning motion causes a lot of motion, resulting in very high temperatures.

Now consider the water on a water slide. In this photo, there is water streaming down the red and yellow lanes. The water looks transparent because there is no churning action. All the water molecules are falling together, at about the same speed. From the perspective of the water molecules, everything is pretty calm. (When you are traveling on a highway, the cars going your direction can seem almost motionless.) Then, when they hit the bottom, that tranquility is shattered and a great deal of commotion ensues, especially if people were riding down the slide along with the water molecules.

Now we will look at our graph again, and include the water analogy. The chromosphere corresponds to the clear water on the water slide. The positive ions going down the voltage drop are experiencing very little motion relative to each other, so the temperature is lower in this area than in the photosphere or corona.

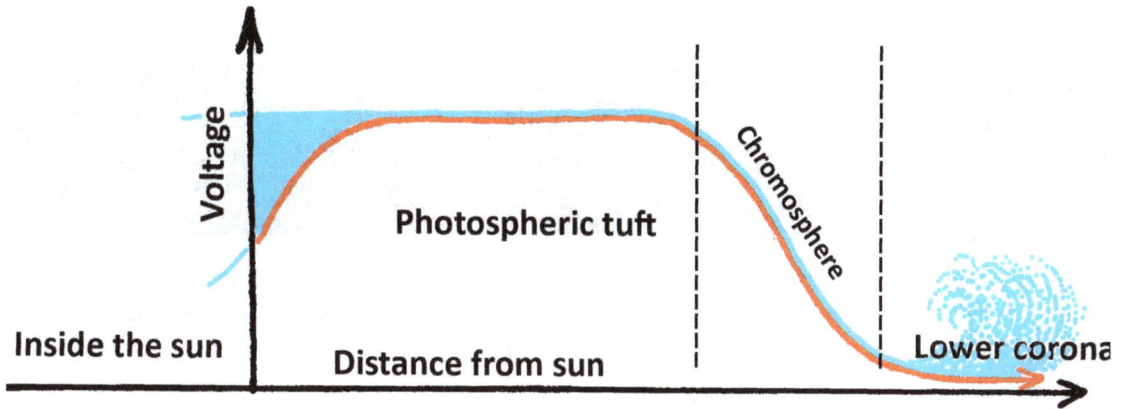

Sun Puzzle 2: Why do we find iron (and other metal ions) in the corona?

First, how do we know that there are iron atoms, and other metal atoms, in the sun?

The discovery of elements in the sun goes back to the late 1800s when the spectrometer was invented. This illustration shows one of these early spectrometers. Notice the triangular glass prism in the center; it was used to break up light into a spectrum of colors. We've all seen prisms produce rainbows from sunlight. It turned out that any kind of light will do this, though not with such a complete spectrum.

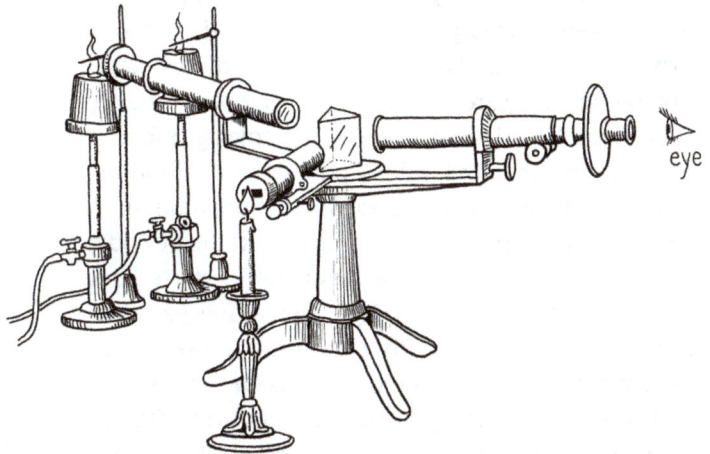

At the back of this antique spectrometer, on the left side, you can see tiny metals sticks that are being heated by gas burners. Various elements (in solution) could be wiped onto these sticks so that the elements could be superheated. The light given off by the flames from these "burning" elements (though they did not burn in the same sense that wood burns) passed through the prisms and could then be analyzed. The scientists quickly figured out that each element produced a unique spectral pattern. These patterns were almost like bar codes for the elements. An element could be identified simply by looking at the spectral pattern. For example, the pattern shown below is for the element neon. No other element has a pattern exactly like this. Real neon signs glow with a very reddish-orange color. The red-orange is so bright that you can't see the yellow, green and blue.

In 1868, a group of scientists turned their spectroscope toward the sun during an eclipse to see if they might be able to identify any of the spectral patterns they had been discovering. To their great surprise, they saw a new pattern! They came to the conclusion that this must be a special element only found in the sun, so they named it "helium," using the Greek word for the sun: "helios." It was years later that someone found helium on the earth, as a decay product of radioactive uranium. This is what the (emission) spectral pattern for helium looks like:

Other spectral patterns observed in the sun (in very small amounts) since that time include: lithium, beryllium, boron, carbon, oxygen, nitrogen, silicon, neon, sodium, calcium, nickel, magnesium, sulfur, and, of course, iron. (Sodium helps to give the sun its yellow glow.)

How did all these elements end up in the sun? One possibility is Stellar Nucleosynthesis. This theory is based on the idea proposed by Arthur Eddington in 1920 that nuclear fusion is the source of energy for all stars including our sun. The Big Bang is believed to have created only hydrogen, helium and possibly lithium; elements larger than these needed another explanation. The Stellar Nucleosynthesis theory was first proposed by Fred Hoyle in 1946, with a refined version published in 1956. In 1957, a small group of scientists joined with Hoyle to publish an addition to the theory in an attempt to explain the creation of elements larger than iron (number 26 on the Periodic Table). The theory states that all elements started out as shown in this chart. Four hydrogen atoms (the red dots at the very top) are fused together to make one helium atom, shown at the bottom as two gray balls and two red balls. The gray balls are neutrons. Oddly enough, protons can turn into neutrons, and vice versa. Electrons are not shown in this diagram. (A positron is a positively charged electron.) This process is believed

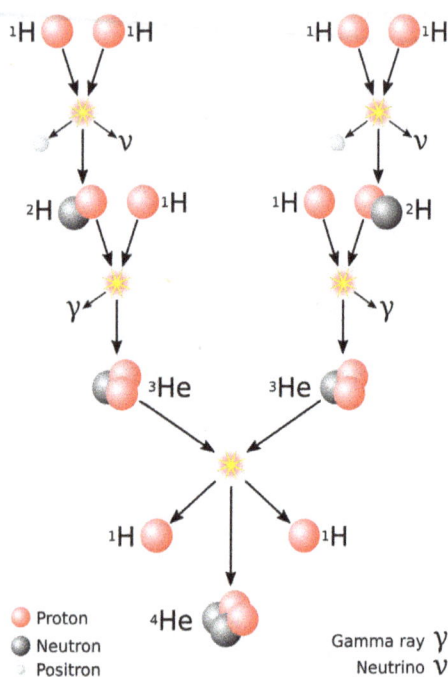

Diagram is from Wikipedia's article on Stellar Nucleosynthesis (public domain image)

to take place in the core of the sun. It is called the proton-proton chain reaction.

A similar process is thought to be able to create elements with 3 to 5 protons (lithium, beryllium, boron). The pressure and temperature inside a medium-sized star like our sun are thought to be sufficient for fusing together these elements. Elements with 6 or more protons, such as carbon, nitrogen and oxygen, would use another fusion process (the CNO process) where extra protons would be added one by one.

Wikipedia confidently states that "the proton–proton chain reaction starts at temperatures of about 4 million degrees, making it the dominant fusion mechanism in smaller stars." For the CNO process. the star's core needs to generate 16 million degrees. Thus, the core of our sun is assumed to be at least 16 million degrees. Our only clue as to what is inside the sun is the sun spots, and they are only 3,000- 4,000° C. Even the very hot corona maxes out at about 2 million degrees. We have no proof at all that any place in or on the sun is 4 million degrees, let alone 16 million. Yet astronomers and nuclear scientists confidently assert that Stellar Nucleosynthesis is correct. All of their proofs are mathematical calculations, not actual observations. Try as they might, scientists still can't make fusion work in a lab setting. (They certainly have been trying. If nuclear labs can figure out how to start and maintain fusion, they will give the world an unlimited supply of "clean" energy.)

If these elements cannot be created in our sun, the Stellar Nucleosynthesis theory offers another option—that they were created in very large stars, or even in supernova explosions, then somehow transported across large distances (a long time ago) where they coalesced, along with hydrogen gas, to form smaller stars. This hypothesis makes several assumptions: 1) There were enough super massive stars capable of filling the universe with smaller stars. 2) It was possible for those newly created heavier elements to be spread out across the universe. 3) Those original massive stars came into existence without the help of previous massive stars. 4) Supernovas are explosions.

The first two assumptions require a great deal of faith in those assumptions. The third assumption defies common sense. The fourth assumption has been called into question by the observation of supernovas "coming back to life" (e.g. Refsdal), and, more recently, a cool star (not expected to explode) that suddenly went supernova (SN2021yfj). Both of these unexpected observations suggest that supernovas might not be explosions in the ordinary sense. If stars are electrical in nature, they might be able to short circuit suddenly, like an explosion in a fusebox, then come back online again if the circuit is reestablished, possibly through Birkeland currents that connect stars and galaxies. If supernovas are not explosions, the fourth assumption is not valid.

There are two other possible sources for these heavier elements. 1) They could have been there all along, as part of the original creation. Oops, this would require a creator—that automatically rules out this possibility for many scientists. 2) These elements are being made either at the very outer edge of the chromosphere (lowest part of the corona), where the protons come crashing down the voltage "slide" and slam into the atoms floating in the sun's atmosphere, or in the plasma filaments at the top of the photosphere where the Z-pinch effect is

happening in arc mode plasma. Lightning is arc mode plasma and it has been observed to create Z-pinches here on earth. The Z-pinch effect strips nuclei of their electrons, then allows all the pieces to rearrange to form new atoms.

In any method of fusion, adding protons to an atom changes its identity, making it into a new element. Neutrons would have to be added, also, but because protons can turn into neutrons, we don't necessarily require a source of neutrons. (In fact, free neutrons are very short-lived; they decay and disappear in a matter of hours or days.) We can't confirm whether fusion is happening in the outer parts of the sun, but this idea should at least be discussed. Why isn't it? Because most astronomers think that nuclear fusion in the core is a fact, not a hypothesis. That's what they were taught.

Regardless of how the iron and other metals got there, they are now flying up into the corona, along with the positive ions of the solar winds. They seem to be part of the ion flow out from the sun.

Summary and conclusions

Despite the fact that modern science has made great progress in its ability to study the sun, there are still basic questions that remain unanswered. Most astronomers and physicists struggle to come up with a reasonable and straightforward explanation for why sun spots are the coolest place on the sun even though they are assumed to be "holes" that should let the heat of nuclear fusion come roaring up from the core. They don't have a good theory for why the corona, the layer farthest from the surface of the sun, has a temperature of millions of degrees. In fact, they don't have a good explanation for why the corona exists at all. They don't know why there are two types of solar wind, and they can't explain how the solar wind could have stopped for two days. Their nuclear synthesis explanation of metals in the sun has many logical problems.

Astronomy was greatly affected by the opinions of Sir Arthur Eddington in the early years of the 20th century, when scientists were discovering the power of the atom and nuclear energy. To him, it was obvious that because the sun was so hot it must be using nuclear fusion to produce its heat. Eddington's idea won the day and it soon became unthinkable to challenge it. During the World War 2 era, interest in nuclear energy was still intense, especially after the discovery of how it could be used to make nuclear bombs. The nuclear fusion theory had gained so much momentum by the late 1900s that it become the only theory taught in schools and universities. It is now assumed to be fact because it is in all the textbooks.

Long before Eddington's fusion hypothesis, Kristian Birkeland was already making discoveries about plasma's role in astronomy. Hannes Alfvén continued with Birkeland's ideas and combined them with plasma research done by Irving Langmuir. Alfvén concluded that plasma physics was the key to understanding not only the auroras, but also the sun, the stars, and even galaxies. In the 1970s, electrical engineer Ralph Juergens realized that since plasma is associated with electrical currents, it was possible that stars were essentially electrical phenomena. He proposed a new model for how the sun works, based on well-established laws of how electricity behaves.

The "electric sun" model can offer reasonable hypotheses for the strange characteristics of the sun that puzzle mainstream astronomers. Sun spots are cool because there isn't any nuclear fusion going on in the core. Sun spots are where a great abundance of protons stream out of the sun unimpeded, creating the slow solar wind. The fast solar wind comes from the anode tufts (granules) that cover the photosphere. The voltage increases as the protons try to "climb" to the top of the tufts, and only a few succeed. When they "fall" down the voltage drop, they fly away from the sun very quickly, creating the fast solar wind. The particles of the solar wind are primarily positive protons, which are repelled by the positive charge of the sun, making them speed up as they stream away from the sun. The solar wind was able to stop for two days because the voltage situation in the tufts changed, perhaps due to the electrical nature of the space through which it was traveling at that time.

The corona is the hottest part of the sun because the protons that are "falling" down the voltage gradient of the anode tufts suddenly hit the atoms in the atmosphere, churning them up and greatly increasing their motion. Heat is molecular motion, so the churning causes heat. Metal atoms in the corona certainly did not come from supernovas. They might have been there from the beginning, or they might be created when the protons come smashing into the lower corona. Metals are perhaps the most enigmatic sun puzzle, but there are excellent reasons to rule out the Stellar Nucleosynthesis hypothesis.

Frequently Asked Questions:

1) Do other planets have auroras? Yes. Auroras have been observed on Jupiter, Uranus and Neptune, as well as on a few of Jupiter's moons.

2) Does the sun act as a source of electrical energy for us here on earth? Not that we know of. Birkeland currents from the sun pass through the poles, but the energy doesn't stay here, it moves on, going toward the outer planets.

3) What is a "coronal hole"? This is a place underneath the corona that is "normal," meaning a place in the photosphere where there aren't any sunspots, just anode tufts. The fast solar wind, produced by the photospheric anode tufts, comes out of the coronal holes. ("Hole" is a bit of a misnomer.)

4) What are spicules? These are "fountains" made of electrons that come up out of the chromosphere then quickly fall back into the sun. The negative electrons are attracted to the positively charged tufts.

5) What about solar system formation?

One aspect of the nebular hypothesis and the plasma hypothesis that is somewhat similar is their claim that the sun and planets of our solar system formed at approximately the same time, as one multi-faceted event. The nebular hypothesis says that the nebular cloud that would eventually form our solar system started out as spinning disc of gas and dust. Most of the mass of this spinning disk would coalesce to form the sun. The outer planets formed where the temperature was cooler and thus lighter gases (such as methane, ammonia and carbon dioxide) could condense. This is proposed as the reason that the outer planets are gaseous.

The plasma explanation also has the planets and sun forming as one event, but for different reasons, based on observations in lab experiments with plasma. The experiments showed that multiple pinches can occur along the same plasma filament. Where the filament is large, a large plasmoid will result. When pinches occur along thinner parts of the filament, smaller plasmoids are formed. They can all form at the same time along the filament. If there are also many different type of atoms present, the elements that are most easily ionized will collect near the center of the filament and those that are not easily ionized will remain at the edges. This is called *Marklund convection*.

Iron is an example of an atom that is easily ionized, meaning that it will easily give up electrons in its outer shells. Therefore, iron would be one of the elements drawn into the center of the filament. Going from the center outwards, you would have iron, nickel, silicon, magnesium, sulfur, carbon, hydrogen, oxygen, nitrogen, helium. Notice that this roughly corresponds to the layers found in the planets of our solar system. This order is also seen in the percentages of these elements in the planets going outward from the sun. Mercury has the largest iron-nickel core for its size. Venus, from what we can tell (though it is a hard planet to study), has a smaller core than Mercury and a greater volume of gas in its atmosphere. Earth has a smaller core than Venus, a larger silicon-based mantle and crust, and an atmosphere of nitrogen and oxygen with water (hydrogen and oxygen) and carbon dioxide (carbon mixed with oxygen). Mars has a smaller core and larger mantle than Earth. The gas giant planets have smaller cores and larger amounts of gas as you continue outward.

Marklund convection does not rule out a creative force forming the planets. It simply gives a physical mechanism consistent with our observations of plasma and atoms.

6) Is plasma astronomy compatible with the idea that a Creator formed the universe? Yes. There is nothing about plasma astronomy that contradicts this idea. Once atoms and subatomic particles came into existence, we can assume that they immediately had their inherent characteristics. Protons were positively charged, neurons were neutral, electrons were negatively charged, and atoms having both a nucleus and orbiting electrons would behave according to their elemental atomic structure (hydrogen would act like hydrogen, helium like helium, etc.). Ions of these elements would behave as ions do. However, elements can't self-organize into the complicated structures and organisms we have here on planet earth.

7) Is the sun really 4.5 billion years old? No, it does not have to be that old. The standard model of astronomy, based on the Big Bang which attempts to use gravity to explain everything, has to assume these long ages in order to make the Nebular Hypothesis and Stellar Nucleosynthesis work. Plasma process can happen very quickly. In the lab, mini-galaxies (such as those seen on page 15) are made in a fraction of a second. In outer space it would take longer, but would not require billions of years.

8) Can we prove that the "life cycle of stars" idea is correct? No, we can't. No one has ever seen a star go through a life cycle. We observe stars of all sizes and colors but the idea that these are stages that all stars go through involves many assumptions and some imaginative thinking. In fact, we have evidence to the contrary. For example:

1) T-type (brown) dwarf stars are much smaller than our sun and have a surface temperature of less than $1,000°$ C. Their spectral patterns indicate the presence of a large amount of methane, similar to Jupiter. These stars do not have enough mass to sustain any fusion at all. According to the standard model, small stars must be at least 75 times more massive than Jupiter (7% of our sun's mass) to have any fusion in their cores. Astronomers sometimes call these dwarfs "failed stars" that do not have enough energy to emit light. Their energy, they say, comes from gravitational collapse. However, in the year 2000, the Chandra x-ray telescope discovered a brown dwarf that is emitting an x-ray flare. The lead researcher said, "We were shocked. We didn't expect to see flaring from such a lightweight object." In the electric sun model, there is no minimum temperature or size requirement. Brown dwarfs are operating near the upper boundary of dark mode plasma (page 11). Any slight increase in the level of current density on its surface will shift the star into normal glow mode. This change will be accompanied by a voltage rise in the plasma of the star's upper atmosphere. This can result in electromagnetic fields that produce a flare of x-rays. This can also happen if a star goes from glow mode to dark mode. (paraphrased from *The Interconnected Cosmos,* by Donald Scott)

2) The star FG Sagittae breaks all the rules of the star evolution story. It is the central star of the planetary nebula He 1-5. This star has changed from blue to yellow since 1955. Recently it has taken a deep dive in its luminosity. The star evolution theory requires long periods of time for a star to change its color or size.

3) Star V Sagittarii (also known as Sakurai's object) was discovered in 1994. It has changed its spectral type and surface composition since it was discovered, not over millions of years.

4) Star V838 Mon, which suddenly appeared in telescopes in 2002, transformed in just a few months from a small star just a little hotter than our sun to a cool, very bright super giant. One astronomer said, "This star has shown a behavior that is not predicted by present theories."

5) Sirius, the "dog star," was recorded in ancient times (by Cicero, Horace, Ptolemy and Seneca) as being red or copper in color. At present it is blue-white. That change did not happen over millions of years.

6) Betelgeuse, the alpha star in Orion, is a red giant. Its diameter has been observed to fluctuate from 480 to 800 million miles, which is 550 to 920 times the diameter of our sun. However, its surface temperature is just a bit over 1,000 C. much cooler than our sun. Some red giants are even cooler. Their density is only one ten-thousandth of earth's atmosphere. They have been jokingly called "red hot vacuums." How can these huge, low density stars maintain hydrogen to helium fusion? The obvious answer is that they can't.

9) Are there other problems with the idea that convection drives heat transport in the sun (in addition to the Reynolds number problem mentioned on pages 20-21)? Yes, here are two more problems: 1) the oscillation problem, and 2) the missing neutrinos problem.

`   1) The sun varies in size and brightness. The size varies by 10 km every 2 hours and 40 minutes. The brightness has a periodicity of a few minutes to one hour. The standard model has no way to explain this, since heat convection from the interior would take many years, perhaps even thousands of year.

        2) When hydrogen atoms combine through fusion to make helium atoms, particles called neutrinos are released. If the sun is powered by nuclear fusion, there should be a massive amount of neutrinos coming out of the sun. In fact, we detect only 1/3 (some say 1/2) the number of expected neutrinos. Instead of going back and challenging their basic assumption (nuclear fusion) they suggest new ideas about neutrinos, such as the possibility that neutrinos might be able to change their form, making them undetectable.

10) Does the electric sun/star model explain pulsars? Yes. There are many problems with the standard model's explanation of pulsars. For decades we have been told that pulsars pulse because they are emitting radiation on just one side and are rotating very quickly. The analogy of a light-house was used to demonstrate how something can appear to be flashing because of its rotation. However, the "flashes" of some pulsars are so fast that the star must be rotating at speeds that break the laws of physics. To get around this problem, astronomers invented the neutron star. Pulsars were said to be made of nothing but neutrons, causing them to be dense enough to be able to rotate at these impossible speeds. The astronomers forgot their basic chemistry lessons. Neutrons can't exist on their own. Outside of an atom that also contains protons and electrons, neutrons decay and disappear in a matter of hours or days. So then they finally realized that a star made of nothing but neutrons was impossible, they invented "strange matter," a fictional type of matter that can only exist if we throw out everything we know about the physics of matter.

        There is a much easier explanation. If stars are electrical in nature, pulsars could easily be a pair of binary stars ("twin" stars that revolve around each other) that are experiencing electrical discharges between them. The characteristics of pulsars are very similar to those of relaxation oscillators used by electrical engineers. The goings-on between the stars could be similar to what happens at a radio station, where electricity is used to generate radio signals. (Pulsars emit radio frequencies of about 600 Hz.) The stars act as capacitors and the plasma between them acts like a non-linear resistor. (For a complete description of how and why pulsars are electrical in nature, you can read chapter 8 in Donald Scott's book, *The Interconnected Cosmos*, listed in the bibliography.)

11) How does the electric sun/star model explain black holes? Black holes probably don't exist. They are likely just plasmoids that have dimmed to the point that they are not emitting any detectable radiation, so they appear dark and empty. The idea that they have extreme gravity that can keep even light from escaping is only supported by theoretical mathematics. Recently, astronomers where surprised to find some black holes emitting x-rays. They immediately had to come up with a rescue device to explain this. The electrical explanation actually expects things like this to happen, since black holes could (like stars) receive a current boost from time to time. The "black hole" at the center of our galaxy started out as the central plasmoid during formation, then dimmed over time.

12) Can the sun really get all its energy from an outside source? Strange as it may seem, calculations show that this is possible. Satellite probes (e.g. Voyager 1) have found a surprising number of electrons floating in empty space. Space is not as empty as we thought. The particles are still sparse, but can reach 10,000,000 electrons per cubic meter. As the sun (and the solar system) moves through space, it would be able to gather enough electrons to keep its energy output going at its current level. The electrons would flow in at the poles. For more details, including the math behind this claim, see Appendix A in Donald Scott's book, *The Interconnected Cosmos*.

13) Does the solar wind ever slow down? Yes. At the outer edges of the solar system, when it goes beyond Pluto, it appears to slow down. It hits a place called the "termination shock" at 75-90 AU (astronomical units—the distance from the sun to earth). Here it runs into the intergalactic medium (a different type of plasma). This slows it down to subsonic speeds. Eventually it reaches the "heliopause" (over 100 AU from the sun) which is the end of the sun's plasmasphere (heliosphere). The particles of the solar wind do not go beyond this point.

14) Has NASA been made aware of these electrical explanations for the properties of the sun? In 2011, Donald Scott, the author of *The Electric Sky* and *The Interconnected Cosmos* gave a lecture (titled "Plasma Physics' Answers to the New Cosmological Questions") to a group of scientists gathered at the Goddard Space Center. You can watch this presentation on the YouTube channel called "suedeslounge." The video has gotten 112,000 views. He also gave a lecture at Case Western Reserve University in 2017 at their school of engineering and computer science about the transistor analogy of the sun's surface. This is posted on YouTube and has received over 37,000 views. The longterm effect of these presentations is still unknown. If there were younger scientists in the crowd, perhaps some of them saw how powerful the electrical explanations are, and will encourage their young colleagues to seriously consider these new ideas. Older scientists probably have too much of their lives invested in the standard model.

15) Are there videos available about the electric sun, for those who prefer watching to reading? Yes. There are dozens of videos about the electric sun, most of them posted by a YouTube channel called "The Thunderbolts Project." You can watch many videos about the information presented in this book. Donald Scott's lectures are posted here, as well as presentations by other ES advocates. The videos are generally short (15-20 minutes) except for the full length Donald Scott lectures which run 45-50 minutes.
DISCLAIMER: The Thunderbolts Project covers more topics than just the Electric Sun. Many people are involved in this project, not just electrical engineers. You might find videos that are a little on the philosophical side, or that blend the mysterious beliefs of ancient cultures with possible astronomical explanations. This book does not endorse anything but the scientific lectures that deal with well-established principles of electricity, magnetism, and plasma science.

Another YouTube channel that advocates for plasma explanations is "See the Pattern."

BIBLIOGRAPHY

BOOKS

Setterfield, Barry J. *Cosmology and the Zero Point Energy*. (Natural Philosophy Alliance Monograph Series, No.1, 2013). Natural Philosophy Alliance, Inc., 2013. ISBN 978-1-304-19508-1

Scott, Donald E. *The Interconnected Cosmos*. Stickmanonstone, Pub., Minneapolis, MN, 2021. ISBN 979-8-9851181-0-0

Scott, Donald E. *The Electric Sky*. Mikamar Publishing. 2006. ISBN 978-0977285112

All three of these books build on the research and writings of the scientists listed on the title page.

WEBSITES and VIDEOS

NOTE: Extensive use of Wikipedia is intentional, since it represents mainstreams ideas and theories.

Temperature of the sun
https://www.astronomy.com/science/how-is-the-temperature-of-the-suns-surface-measured-through-its-much-hotter-atmosphere-the-corona/
https://www.space.com/17137-how-hot-is-the-sun.html
https://science.nasa.gov/sun/facts/
https://www.astronomy.com/science/how-is-the-temperature-of-the-suns-surface-measured-through-its-much-hotter-atmosphere-the-corona/

Corona
https://spaceplace.nasa.gov/sun-corona/en/
https://en.wikipedia.org/wiki/Stellar_corona
https://phys.org/news/2017-08-clue-mystery-sun-hot-atmosphere.html (quote on page 5)
https://www.space.com/solar-waves-sun-corona-mysteriously-hotter-than-surface

Solar wind
https://en.wikipedia.org/wiki/Solar_wind
https://news.uchicago.edu/explainer/what-is-solar-wind
https://www.techexplorist.com/solar-wind-evolves-increasing-distance-sun/28117/
https://scied.ucar.edu/learning-zone/sun-space-weather/solar-wind
https://www.aeronomie.be/en/encyclopedia/solar-wind-speeds-fast-and-slow
https://news.mit.edu/1999/solarwind-1215 (solar wind shut off in 1999)

Sun spots
https://science.nasa.gov/sun/sunspots/
Also many of Donald Scott's video lectures

Birkeland currents
https://en.wikipedia.org/wiki/Kristian_Birkeland#Research

Hannes Alfvén
https://en.wikipedia.org/wiki/Magnetohydrodynamics
https://en.wikipedia.org/wiki/Hannes_Alfv%C3%A9n#Research
https://en.wikipedia.org/wiki/Van_Allen_radiation_belt

Irving Langmuir
https://en.wikipedia.org/wiki/Irving_Langmuir
https://www.britannica.com/biography/Irving-Langmuir
Video intro to Electric Sun by Donald Scott, presented at EU-UK 2018:
          https://www.youtube.com/watch?v=eHi6ib9STPg&t=4s

Double layers
https://www.plasma-universe.com/double-layer/

Quote about engineers
https://en.wikipedia.org/wiki/Talk%3APlasma_cosmology%2FArchive_11

Ralph Juergens
https://www.velikovsky.info/electric-sun-model/

Birkeland currents
https://www.plasma-universe.com/birkeland-current/

Nebular Hypothesis
https://geo.libretexts.org/Bookshelves/Geology/Book%3A_An_Introduction_to_Geology_(Johnson_
    Affolter_Inkenbrandt_and_Mosher)/08%3A_Earth_History/8.02%3A_Origin_of_the_Solar_SystemThe_
    Nebular_Hypothesis

Photosphere
https://en.wikipedia.org/wiki/Solar_granule
Voltage: https://en.wikipedia.org/wiki/Voltage

Composition of sun
http://hyperphysics.phy-astr.gsu.edu/hbase/Tables/suncomp.html

Stellar Nucleosynthsis
https://en.wikipedia.org/wiki/Stellar_nucleosynthesis

Reappearing supernova
https://www.livescience.com/space/cosmology/scientists-watched-a-reappearing-supernova-explode-
    5-times-in-a-row-and-it-could-help-reveal-how-fast-the-universe-is-expanding

Cool supernova
https://www.sciencedaily.com/releases/2025/08/250821004234.htm

Brown dwarf
https://iopscience.iop.org/article/10.1086/312817

Heliopause
https://theplanets.org/the-heliopause/
https://en.wikipedia.org/wiki/Heliosphere#Termination_shock

2.17.26